FUN FACTS FOR KIDS

DINOSAURS

CONTENTS

INTRODUCTION

Hey there, young explorer!

Have you ever wondered what it would be like to take a stroll with a mighty Tyrannosaurus Rex or play hide-and-seek with a swift Velociraptor? Maybe you've imagined listening to the gentle songs of a herd of Brachiosaurus as they munch on treetops. Welcome to the world of "Fun Facts for Kids - Dinosaurs"!

Long before humans made our grand entrance, our planet was ruled by these incredible creatures we call dinosaurs. From the towering giants that shook the ground with every step to the nimble bird-like dinos that darted through the skies, these magnificent beasts have captured the imaginations of kids of all ages for generations.

In this book, we'll embark on a thrilling journey back in time to discover some of the most awe-inspiring, jaw-dropping, and, sometimes, just downright quirky facts about these prehistoric titans. Whether you're a budding paleontologist or simply curious about what made these creatures so special, you're in for a treat.

So, strap on your adventure boots, grab your favorite dino toy, and let's set out on an epic adventure to a world lost in time, where dinosaurs roamed freely.

Ready? Let's dig in!

FUN FACTS

1. Tyrannosaurus Rex Teeth: The Tyrannosaurus Rex, often called T-Rex, had teeth that were up to 12 inches long. That's about the size of a ruler! You wouldn't want to meet up with a T-Rex, would you?

2. Dinosaurs: They are an ancient group of reptiles that roamed our planet millions of years ago. Some were as tall as buildings, while others were as small as birds. Each one had its own unique features and sounds! Can you name any of the dinosaurs in the picture?

3. Dinosaur Names: If you combine the words from the Greek words "deinos" meaning terrible and "sauros", which means reptile or lizard you have the word Dinosaurs. When put together, it means "terrible lizard." A pretty fitting name, don't you think?

4. Feathered Friends: Some dinosaurs, like the Velociraptor, had feathers. It's believed that birds evolved from these feathered dinosaurs. The first dinosaur fossils that appeared to have feather-like features were found in the 1990s. Not so long ago, was it?

5. Who Named Dinosaurs: The term Dinosauria: Sir Richard Owen, in the year 1842, first came up with the name for these ancient animals. Before that, men had no concept of anything that resembled a dinosaur. People didn't understand what dinosaurs were even if they found their skeletons. Sir Richard Owen's career lasted 60 years. He was born in 1804 and died in 1892.

6. Herbivores: Not all dinosaurs ate meat. Many were herbivores, which means they ate plants, while others were carnivores and hunted for food. Scientists can tell which dinosaurs were herbivores (plant eaters) by looking at the kind of teeth they had. They had to be able to grind up their food.

This is as Brachiosaurus eating from an araucaria tree.
By Davide Bonadonna, Milan, Italy.

7. Carnivores: The Carnivores (meat eaters) probably ate anything they could catch and had sharp teeth. Some of them may even have hunted in packs.

8. Biggest: One of the most giant dinosaurs was Argentinosaurus, which could grow over 100 feet long was found in Argentina in 1987. That is as long as three school buses lined up end to end.

9. Smallest: The Microraptor was only the size of a crow, weighing about 2 pounds! It appears to have had feathers and four wings. Their wings were more likely used for gliding and not flying.

An artist's illustration of *Microraptor* with iridescent plumage (Jason Brougham/University of Texas)

10. Cold-Blooded or Warm-Blooded?: Scientists are still debating whether dinosaurs were cold-blooded like reptiles or warm-blooded like mammals and birds. The T-rex and Allosaurus were probably warm-blooded dinosaurs, and the Triceratops and Stegosaurus were cold-blooded like a snake! Oh, I don't like snakes, do you?

11. Dinosaur Eggs: Many dinosaurs laid eggs, some of which have been found preserved. They come in various sizes, from tiny ones to those as big as a football. That football-sized egg would make one big omelet! It could probably feed a whole family.

12. Dino Sounds: No one knows exactly what sound dinosaurs made, but they probably communicated with each other through grunts or even chirps. Some might have roared, and others may have sounded like a horn. No one on Earth has ever heard a dinosaur sound, but their skeletons may provide some hints!

OK young dino detective, what do you think these two sounded like?

13. Dinosaur Lifespan: Some dinosaurs could live to be quite old. For example, it's believed that a T-Rex could live up to 30 years. For kids that seems really old but for us adults it is not very long at all, ask your parents.

14. Dino Footprints: In some places around the world, you can still see dinosaur footprints in the rock. These fossils give clues about their behavior and environment. These footprints were left in mud before it turned to stone.

15. Flying Dinosaurs: Technically, flying 'dinosaurs' like Pterodactyls aren't dinosaurs at all. They are pterosaurs, a different group of ancient reptiles. Some of these pterosaurs were small enough to fit in your hand, and others were so big that their wings, from tip to tip, were up to 30 feet wide. Wow, I'd hate to see one of those over my head.

16. Dino Defense: Some dinosaurs, like the Stegosaurus, had sharp spikes on their tails to defend themselves from predators. Other dinosaurs protect themselves by staying together in a herd, running away, using camouflage, and using teeth, claws, or horns as weapons.

17. Three Main Periods: Dinosaurs lived during the Mesozoic Era. This Era is separated into three main periods: Triassic, Jurassic, and Cretaceous. Because of the popularity of the movies, I'm sure you've heard about the Jurassic period.

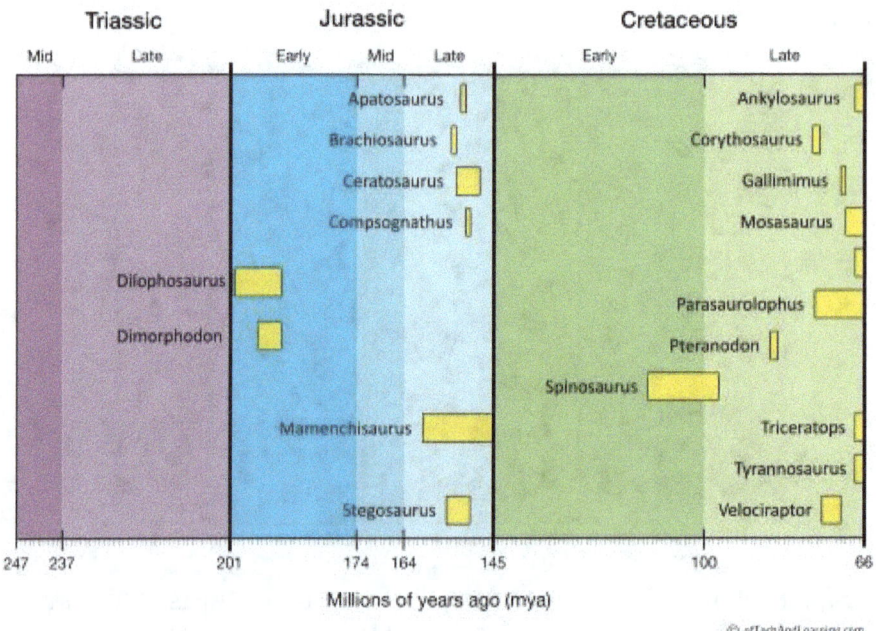

18. Comet or Asteroid Impact: It's believed that the extinction of the dinosaurs was partly caused by a massive comet or asteroid hitting Earth about 65 million years ago.

An artist's impression of what an asteroid colliding with Earth might look like. Sixty-six million years ago an event like this, although on a much smaller scale, caused 75% of all animals to die out. Image: Don Davis Via **NASA Image and Video Library** ⊠

19. Herding Behavior: Some plant-eating dinosaurs, like the Triceratops, might have traveled in herds, similar to today's animals like elephants or cows do now.

20. Dinosaur Colors: We don't know for sure what color dinosaurs were. Many artists imagine them in greens and browns, but they could have been brightly colored, too! Since no one has ever seen a dinosaur, we have to guess. What color do you think dinosaurs were?

Dilophosaurus

21. Dino Brain Size: The Stegosaurus had a brain the size of a walnut, which is very small considering its huge body size! It takes 35 walnuts to make a pound. Your brain weighs somewhere around 3 lbs. That would be a lot of walnuts.

22. Dinosaur Eyelids: Some dinosaurs, like the Velociraptor, had a third eyelid to help protect and moisten the eye. Some archeologists think they were related to ancient birds. If they were like birds, they would have had this third transparent eyelid to moisten and keep the eye clean.

23. Longest Name: The dinosaur with the longest name is Micropachycephalosaurus, which means "tiny thick-headed lizard." It was a herbivore and lived in the area we know as China. It was a tiny dinosaur with a really long name. Can you say Micropachycephalosaurus? I can barely say it myself.

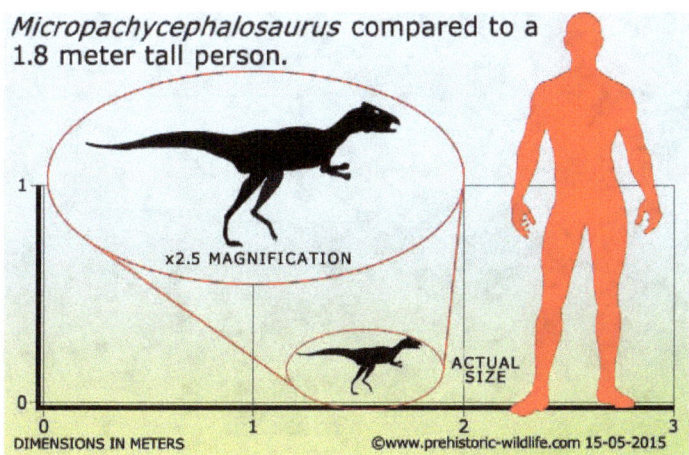

Micropachycephalosaurus compared to a 1.8 meter tall person.

x2.5 MAGNIFICATION

ACTUAL SIZE

1

0

0   1   2   3

DIMENSIONS IN METERS   ©www.prehistoric-wildlife.com 15-05-2015

24. Swimming Dinosaurs: Not all dinosaurs lived on land. The Spinosaurus is believed to have been a good swimmer and might have spent much time in the water. I bet you like swimming and playing in the water at the pool or the beach. I know I like the beach.

Reconstructed Spinosaurus skeleton at the Hong Kong Science Center. Photo by Hong Kong Tourism Board

25. Dino Nests: Some dinosaurs laid eggs, just like birds do today. Some ancient dinosaurs built nests for their eggs, similar to birds and crocodiles. They would then guard and take care of their young.

26. Dino Eggs: Studies show that dinosaur eggs are colorful. Eggs from different species could have had different colors, thicknesses, and other features depending on the environment in which they were laid.- Wikimedia Commons provides this information.

27. Sleeping Dinos: Scientists think some dinosaurs, like modern birds, slept with their heads tucked under their arms. A fossil of a sleeping dinosaur was found in China. Sediment or mud covered the animal while it was asleep, burying the animal, and it turned into a fossil.

28: Jurassic Mammals: During the time of the dinosaurs, there were also small mammals scurrying around. These tiny creatures would eventually evolve and give rise to the variety of mammals we see today, including everything from little mice to the great big elephant.

29. Brachiosaurus: Scientists first believed that the Brachiosaurus lived in water, but they lived on land. Its most distinct feature is its long neck. It weighed up to 28 tons and was up to 85 feet long from head to the tip of its tail.

30. Horned Dinosaurs: The Triceratops is one example of a horned dinosaur. It has three horns, one on its nose and two further back on its head. They were probably used to fight off predators and protect themselves.

31. Lifespans: Dinosaurs lived on Earth for over 160 million years. The oldest T-Rex lived for over 35 years. Scientists can tell how long by counting the number of growth rings in their bones, like the rings in a tree can tell its age.

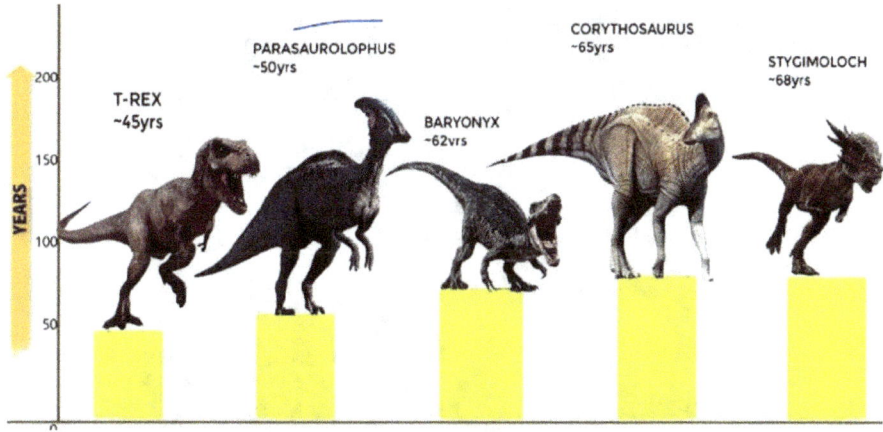

32. Fossil: Fossils are animal and organism parts that have been naturally preserved. Most fossils are bones, teeth, wood, or shells, but sometimes even nests or footprints are also considered fossils.

33. Hollow Bones: Dinosaurs related to birds, like the T-Rex, had hollow bones. This made them lighter and possibly faster. Birds also have hollow bones. This is one reason scientists believe dinosaurs are related to the birds you see today.

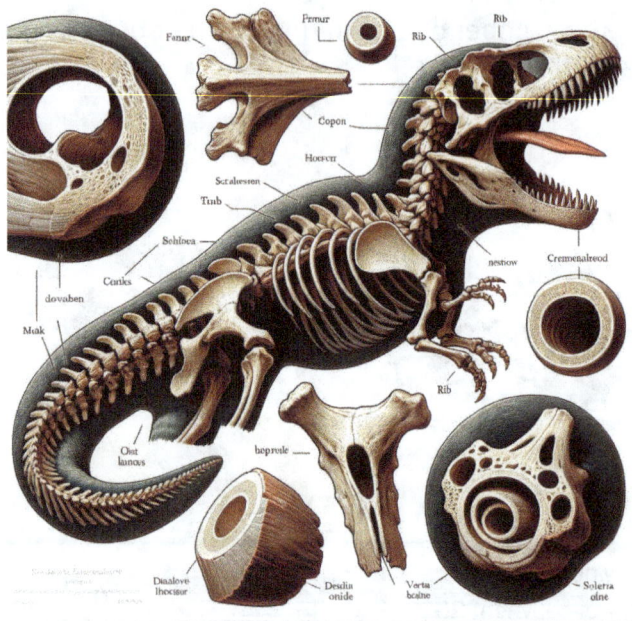

34. Tyrannosaurus Rex: The most well known T-Rex lives in the Chicago Museum of Natural History and has been nicknamed Sue. Sue was found in 1990 in South Dakota. It is not known if Sue was a male or femaled

35. T-Rex teeth: Tyrannosaurus rex had teeth as long as bananas. The skull was around 5 feet long. Museums around the world display T-Rex skulls, both original fossils and replicas. These exhibits help scientists and the public alike to understand and appreciate the scale and power of this prehistoric predator.

36. Velociraptor: The name Velociraptor means "swift seizer." Velociraptors in the movies look as big as a man, but they were only about the size of a turkey! They hunted in groups, so getting away from them would still be very scary! I think it would be really, really fast!

Velociraptor reconstitution

Scientific reconstitution   Jurassic park reconstitution

37. Hesperornis: This bird was flightless but swam in the ocean, catching fish. It would look a bit like a dinosaur mixed with a loon. It lived during the Cretaceous era and had a long neck, waterproof feathers like a duck, and webbed feet.

38. Paleontologists: They use tools like brushes and chisels to uncover dinosaur bones carefully. A scientist who studies the history of the earth by studying fossils is a Paleontologist. If digging in the dirt sounds like fun to you, you may grow up to be a Paleontologist!

39. Lambeosaurus: This duck-billed dinosaur is known for its hatchet-shaped hollow bony crest on its head. It was first discovered in what is now known as Canada in 1914. It was a plant eater or herbivore.

LAMBEOSAURUS

1 metre
3 feet

Lambeosaurus (LAM-bee-oh-SORE-us)
"Lambe's lizard"
Period: Late Cretaceous
Length: 30 feet (9 metres)
Location: North America

A robust herbivore with a distinctive head crest.

© 2010 Encyclopædia Britannica

40. Brachiosaurus: Brachiosaurus had nostrils on top of its head. They are heavier and much bigger than the Brontosaurus, but both are very big. These massive animals were 85 feet long from the tip of their snouts to the end of their tales.

41. Spinosaurus: It had a long spine that looked like a sail on its back. It was carnivorous and had smooth cone-shaped teeth. They had short legs and probably spent most of it's time living in and around water. It had a narrow snout that was ideal for catching and eating fish.

42. Dinosaur footprints: The footprints of dinosaurs are called "track-ways" and have been found on every continent on Earth. By studying these tracks, scientists can estimate an animal's weight, speed, and how it moved. Have you left any footprints in the mud? A scientist in the future might look at them and know more about you. WOW!

43. Ankylosaurus: It was like a living tank with a club tail. The name means "fused lizard" because it had huge plates of bone in its skin. The plates covered its back sides and even its eyelids and served as protection. It was big, about 20 to 30 feet long, and weighed 4- 8 tons.

44. Iguanodon: The Iguanodon had a thumb spike that it possibly used for defense. The animal probably spent time grazing while moving about on four legs, although it could walk on two. The teeth were ridged and formed sloping surfaces whose grinding action could pulverize its diet of low-growing ferns that grew near streams and rivers.

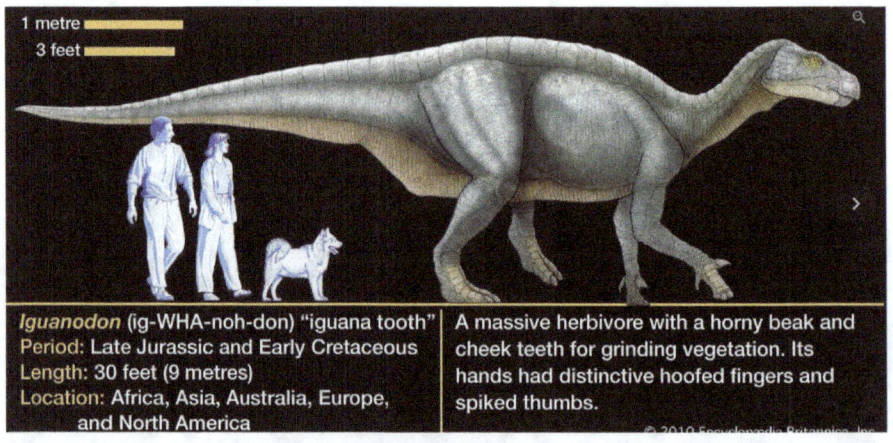

1 metre
3 feet

Iguanodon (ig-WHA-noh-don) "iguana tooth"
Period: Late Jurassic and Early Cretaceous
Length: 30 feet (9 metres)
Location: Africa, Asia, Australia, Europe,
        and North America

A massive herbivore with a horny beak and cheek teeth for grinding vegetation. Its hands had distinctive hoofed fingers and spiked thumbs.

© 2010 Encyclopædia Britannica, Inc.

45. Dinosaur Environment: Dinosaurs lived in various habitats, from deserts to forests. Some remains, such as footprints, indicate where dinosaurs actually lived. Their bones tell us only where they died if they have not been scattered or washed far from their place of death.

46. Coprolite: Fossilized dino poop is called coprolite and tells us about their diet.(25 inches x 6 inches) This coprolite is most likely from a T-Rex. An x-ray shows a high percentage of crushed bone pieces. It probably crushes as the T-Rex ate its prey. That is one big poop!

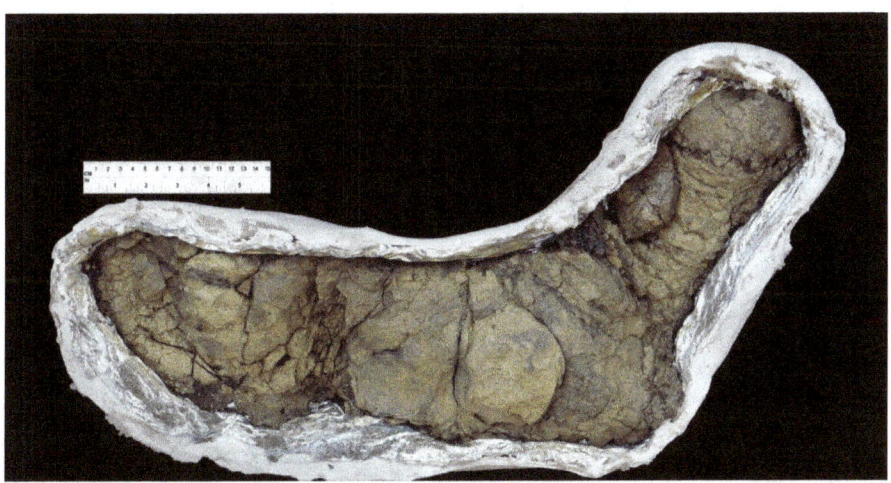

47. Pachycephalosaurus: This dinosaur grew to be about 16 feet long and weighed around 1,000 pounds. The hard dome on it's head might have been used in head-butting contests, similar to modern-day rams.

48. Deinonychus: It had a large, sickle-shaped claw on its foot. Some researchers think this dinosaur was a fast-moving hunter, but others believe it may have been slower and more agile than first imagined. The claws on its feet helped attack or that it allowed the animal to climb or pin smaller prey to the ground.

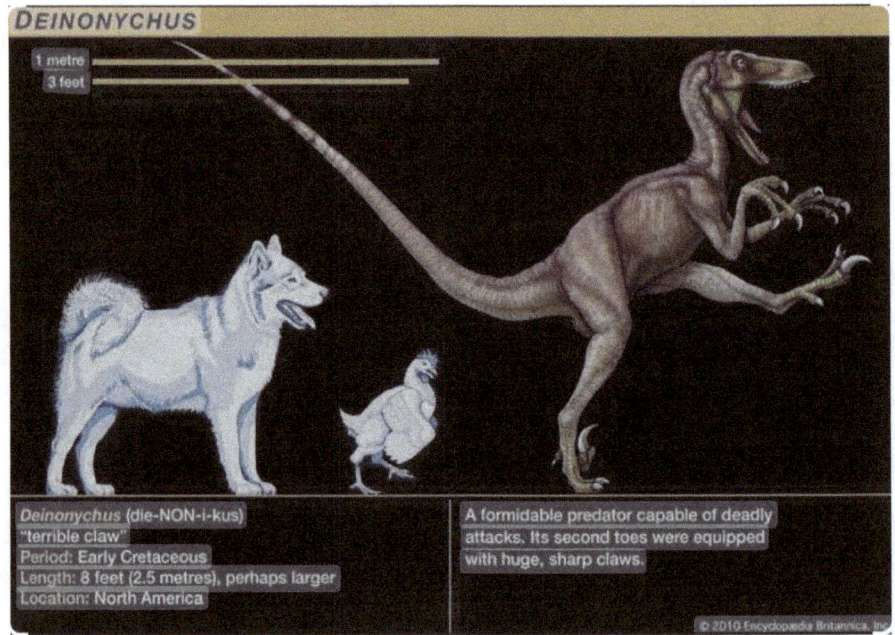

DEINONYCHUS

1 metre
3 feet

Deinonychus (die-NON-i-kus)
"terrible claw"
Period: Early Cretaceous
Length: 8 feet (2.5 metres), perhaps larger
Location: North America

A formidable predator capable of deadly attacks. Its second toes were equipped with huge, sharp claws.

© 2010 Encyclopædia Britannica, Inc.

49. Dinosaur Migration: Some dinosaurs migrated to find food. Their teeth fossils tell the tale that they migrated like elephants and birds do now. Stomach stones in long-necked dinosaurs traveled more than 600 miles in the dinosaurs' tummy.

50. Ears: Dinosaurs did not have external ears like we do. They had inner ears and very good hearing. It is believed the dinosaur's ear was very much like that of a crocodile's. When scientists CT scanned a T-rex skull, they found an inner ear cavity similar to that of modern reptiles.

51. The Troodontid: It may have been one of the most intelligent dinosaurs. It was thought to be a lizard when first discovered, but later fossils show that it is more of a birdlike dinosaur.

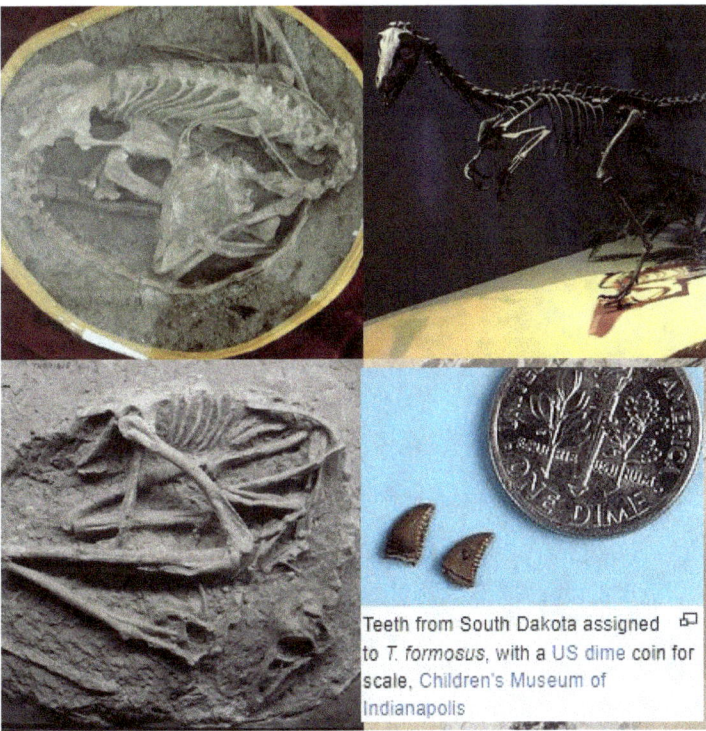

Teeth from South Dakota assigned to *T. formosus*, with a US dime coin for scale, Children's Museum of Indianapolis

52. Raptors: The word raptor refers to a group of small to medium-sized, feathered dinosaurs known as dromaeosaurids. The presence of feathers was revealed through fossil evidence. Raptors were among the most intelligent dinosaurs. Their brain-to-body size ratio was higher than most reptiles, suggesting a greater level of cognitive skills.

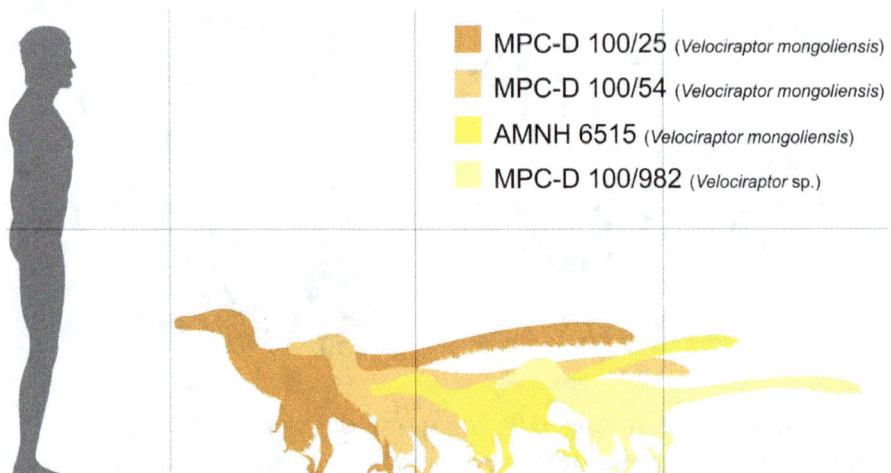

MPC-D 100/25 (*Velociraptor mongoliensis*)
MPC-D 100/54 (*Velociraptor mongoliensis*)
AMNH 6515 (*Velociraptor mongoliensis*)
MPC-D 100/982 (*Velociraptor* sp.)

Velociraptor (top) and Utahraptor (bottom) specimens compared to a 1.8 m (5.9 ft) tall human

A

5 cm

B

C

D

E

F

Uses of the claw

53. Apatosaurus: Sauropods, like the Apatosaurus, had long tails they could whip. It was a large, long-necked animal that walked on all four legs. Apatosaurus means "deceptive lizard". They were 69 to 75 feet long.

54. Theropods: They had hollow bones, three toes, and claws on each limb; some may have had feathers. They were two-legged, primarily carnivorous dinosaurs, but further study has revealed a few that are herbivores and insectivores. That's a new word "insectivores", can you guess what an insectivore eats?

Theropod size comparisons

55. Therizinosaurus: They had the longest claws of any land animal in history. They were almost three feet long! I wouldn't want to trim or paint those. This herbivore had feathers and weighed 6,600 to 11,000 pounds. It was 13-16 feet high and 30-33 feet long.

56. Gastroliths: Some herbivorous dinosaurs, like the Ankylosaurus, ate stones to help digest plant material. These stones are called gastroliths. They might provide significant new information and insights into the lives and behavior of dinosaurs.

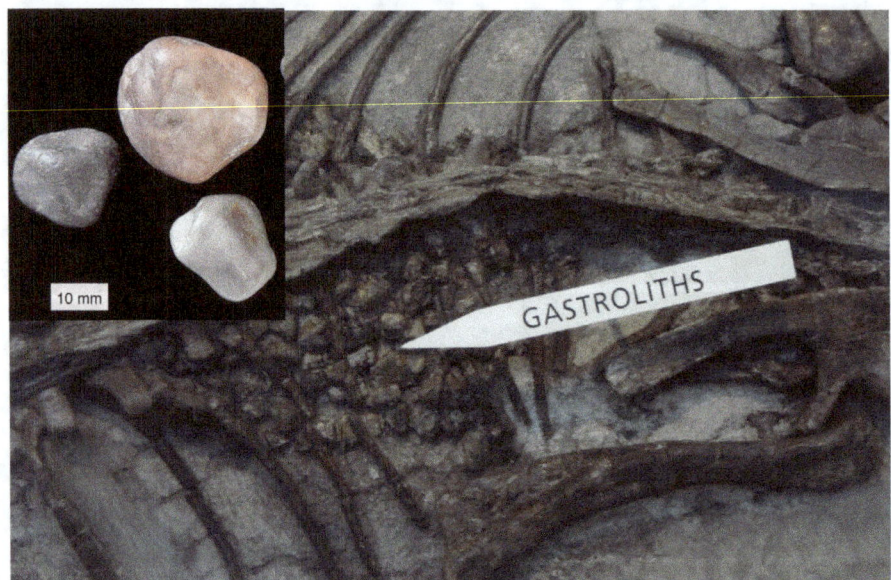

Here we see the stones in the stomach region of this fossil.

57. Hatchlings: Some baby dinosaurs could walk as soon as they were hatched, and others could not. Sauropod dinosaurs, the largest animals ever, could walk when hatched. This includes Titanosaurs and the Mosasurus that walked on four legs when born and later on walked on two.

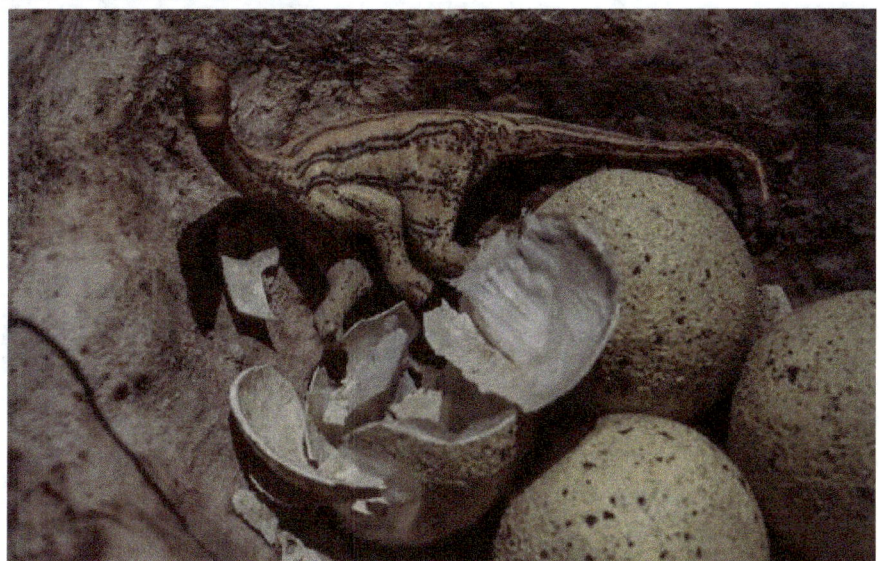

58. Tiny Arms: Some dinosaurs, like the Carnotaurus, had tiny arms even shorter than T. rex's. They also had a horn above each eye. The Carnotaraurus walked on two legs and stood about 29 feet tall. It weighed around 3309 pounds.

59. Protoceratops: This dinosaur, discovered in 1923 and 2001, looked a lot like a modern-day rhino without a horn. It is thought to have been an ancestor of the Triceratops. It was only about 6 feet long. The fossil was discovered in what is now known as Asia.

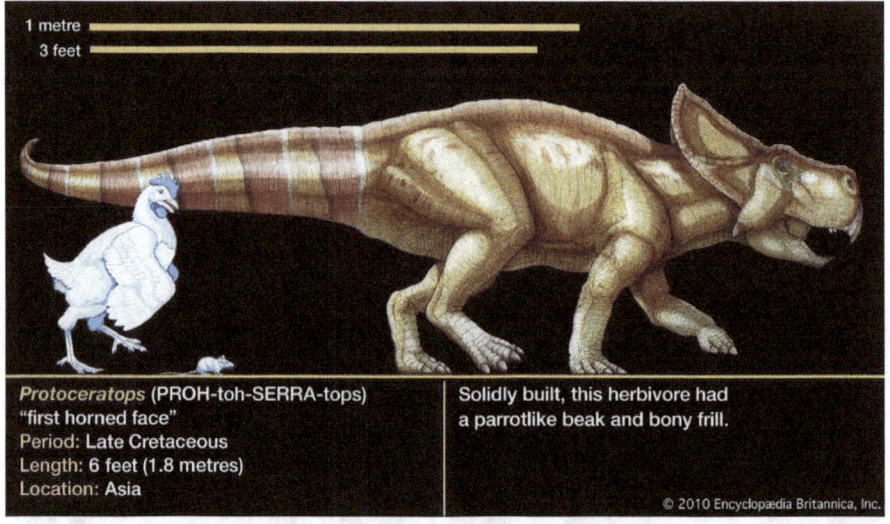

1 metre
3 feet

Protoceratops (PROH-toh-SERRA-tops)
"first horned face"
Period: Late Cretaceous
Length: 6 feet (1.8 metres)
Location: Asia

Solidly built, this herbivore had a parrotlike beak and bony frill.

© 2010 Encyclopædia Britannica, Inc.

60. Sense of smell: Several dinosaurs likely had a great sense of smell. The T-Rex comes up again and probably had one of the best smell detectors! This helped a lot when hunting. Scientists also think that the herbivore, Erlikosaurus had a good sense of smell.

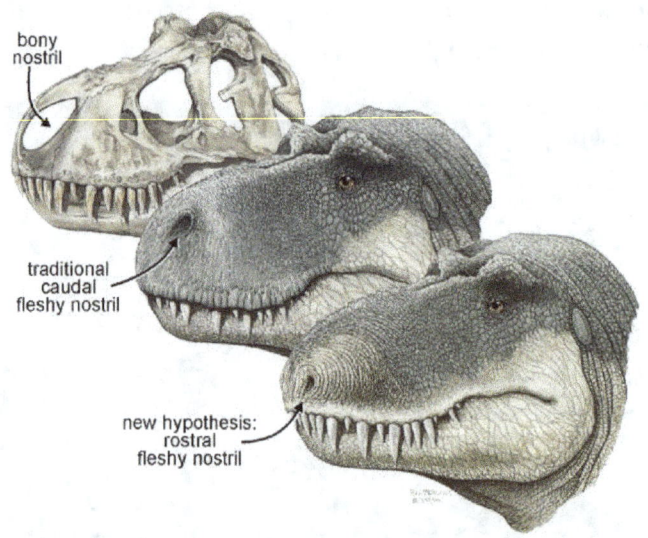

bony
nostril

traditional
caudal
fleshy nostril

new hypothesis:
rostral
fleshy nostril

Changing nostril position in *Tyrannosaurus rex*. The middle image is the traditional view with the nostril located more to the rear of the head. The image at lower right is a new restoration based on Witmer's study reflecting the forward position of the nostril. The image at upper left is the skull.

Copyright *Science*/Paintings by W. L. Parsons under the direction of L. M. Witmer

61. Not a Dinosaur: The Mosasaurus wasn't a dinosaur but a huge marine reptile, reaching a length of about 42-59 feet long, longer than the Tyrannosaurus Rex. Its top speed was around 30 miles per hour, and it weighed up to 78 tons and lived mainly on a diet of sea turtles. It lived in shallow coastal waters and lagoons. I would never want to swim with a Mosasaurus, would you?

62. Spinosaurus: The biggest carnivorous dinosaur was the Spinosaurus, weighing 13 to 22 tons and 46 to 59 feet long; it is bigger than the Tyrannosaurus. The name Spinosaurus means "spined reptile". It has a long, narrow skull about 6 feet long that resembles a crocodile with nostrils near the eyes, not at the end of the snout. It spent lots of time in the water feeding on fish.

63. Ceratopsians: Herbivorous dinosaurs like the Triceratops had a beak. These animals could be as small as a dog or bigger than a car. Not only did they have beaks like parrots, they had hips like a bird. They could snip off plants and tree branches to grind with their powerful teeth. They moved in herds and lived in forested areas.

64. Body Armor: The Gastonia dinosaur had body armor made of bony plates. It is a plant eater or herbivore. The large spikes covering the body and tail helped protect against predators. It had a considerable length of 16 feet and weighed around 4,200 pounds.

65. Why Migrate: Some dinosaurs were massive creatures and required a lot of food. They migrated to find more abundant food during seasonal weather changes when climates and conditions are too hot or too cold. Some dinosaurs might have migrated for breeding, which could ensure the survival of their species by lessoning the risk of predation on eggs and young dinosaurs.

66. Baryonyx: Unlike most theropods, Baryonyx ate fish. It is a carnivore with sharp, finely serrated teeth. It feeds on fish and Iguanodon. Baryonx means "heavy claw". The animal has a very large claw on the first finger.

67. Sauropods: Dinosaurs like the Brontosaurus and Titanosaurus had very long necks and long tails and were big, very big. They were so big when paleontologists (Do you remember that word?)found the fossils, they believed they were fossilized trees. This family of dinosaurs was the largest land animals in history. They were plant eaters grazing on treetops.

,000 kg 1 square		
Sauroposeidon	28 m, 50-60	
Supersaurus	34 m, 35-40	
Apatosaurus	26-30 m, 32-72	
Ankylosaurus	6-8 m, 5-8	
Stegosaurus	8-9 m, 3-7	
Spinosaurus	14-18 m, 7-20	
Shantungosaurus	15 m, 13-16	
Tyrannosaurus	12.3-13 m, 8-14	
Triceratops	8-9 m, 6-12	
Argentinosaurus	36-40 m, 77-100 t	
Puertasaurus	30-35 m, 80-100 t	
Patagotitan	37 m, 69-77 t	
Alamosaurus	30 m, 44-88 t	

https://en.m.wikipedia.org/wiki/File:Biggest-Dinosaurs-ver19-en.svg

68. Egg Theif: Oviraptors did not really steal eggs, despite their name meaning "egg thief." It was a strange-looking toothless dinosaur. When a fossil egg matching the "stolen" egg was found with a tiny oviraptorid in it, they knew that the egg was not stolen but was being protected by its parent.

Actual fossil found.

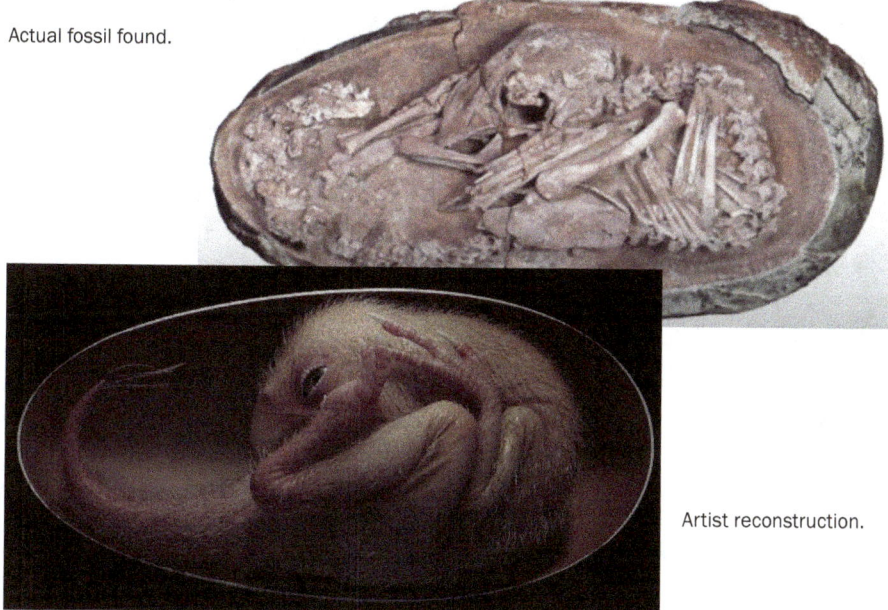

Artist reconstruction.

69. Bite Strength: The T. rex could bite with a force of 12,800 pounds. When feeding on its prey, this bite could actually crush bones. I certainly am glad I will never be a meal for this beast. What about you? We are small, so we would actually be a snack, not a meal. Don't you think?

The Tyrannosaurus rex known as Stan, excavated in South Dakota in 1992, is one of the most complete Tyrannosaurus rex skeletons in the world. Greg Latza / AP Images

70. Duck Bills: The Hadrosaurs were duck-billed dinosaurs. They had hundreds of teeth. They probably dined on plants close to the ground, mostly leaves. In addition to the duck-billed snout, some Hadrosaurs had large crests on their heads used to display and/or make noises.

Maiasaurus nest.

71. Pentaceratops: This animal was a herbivore closely related to Triceratops. It had a very large skull that was about 7.5 feet. They grew to around 18 feet in length. The frill is much longer than that on Triceratops and points up and back. The brow horns curve forward and are very long.

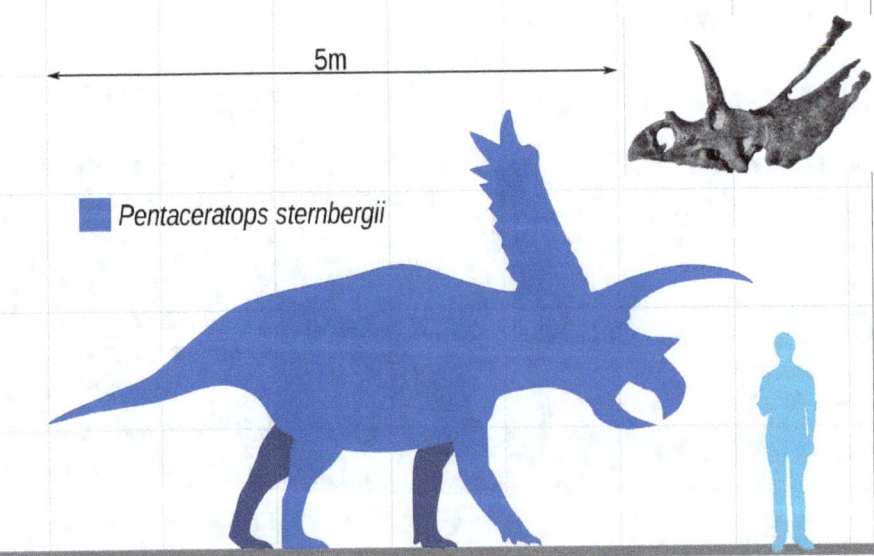

5m

Pentaceratops sternbergii

72. Carnivorous dinosaurs: These meat-eating dinosaurs had sharp claws to catch their prey. They include the T-Rex and the even larger Spinosaurus. It is believed that these theropods were the only dinosaurs to get continuously smaller.

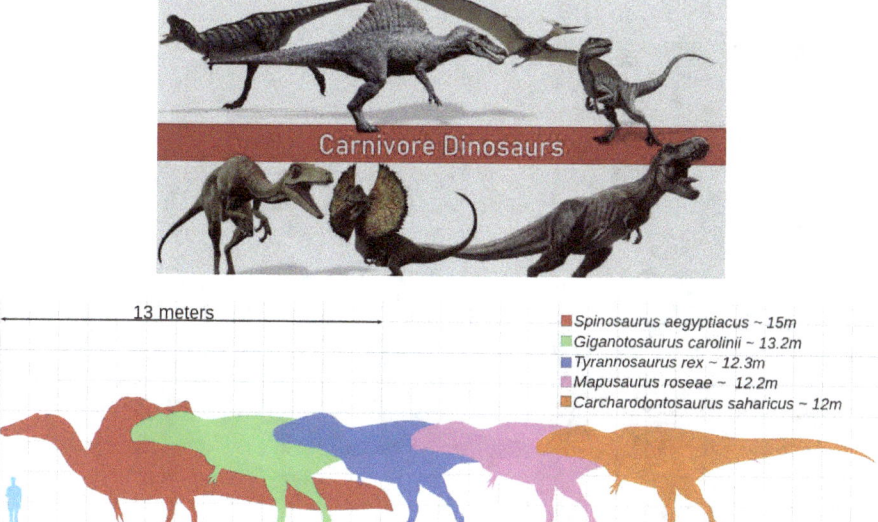

73. Fossils: Dinosaur bones have been found on all seven continents, including Antarctica. The most dinosaur fossils and the greatest variety of species were found in the deserts and badlands of North America, China, and Argentina.

74. Dinosaur Teeth: Scientists have studied these teeth since 1822, when an Iguanodon tooth was found in Sussex, England. Types of tissues, tooth wear, replacement patterns, and how they are attached have been discovered about many. They come in all shapes and sizes, adapted to the way they are used.

Tyrannosaurus Rex tooth.

Teeth from South Dakota assigned to *T. formosus*, with a US dime coin for scale, Children's Museum of Indianapolis

75. Relatives: Dinosaurs' closest living relatives today are birds and crocodiles.

76. Dromaeosaurus: It had sharp, serrated teeth. Agile, lightly built and fast-running they were among the best preditors. The second toe of each foot was flexible and had a killing claw or talon.

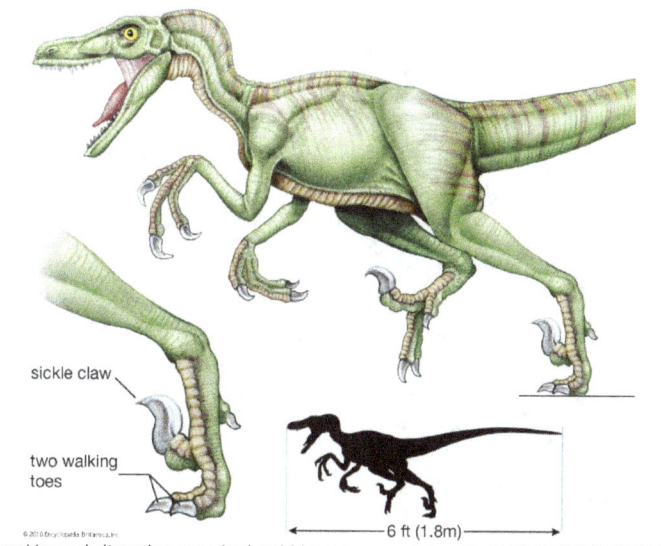

sickle claw

two walking toes

6 ft (1.8m)

https://www.britannica.com/animal/dromaeosaur#/media/1/171765/148664

Access Date November 13, 2023

77. Lion of the Jurassic: The Allosaurus was known as the "lion of the Jurassic." It weighed two tons and was about 35 feet or longer. The forelimbs were larger than Tyrannosauruss, with three fingers ending in sharp claws.

78. Speed: Some dinosaurs were capable of reaching speeds comparable to today's cheetahs. To put this in perspective, consider the speed of a cheetah that accelerates from 0-60 miles per hour in less than 3 seconds, similar to a fast car's acceleration.

79. Very Big Raptor: The Utahraptor was the largest raptor, even bigger than the Velociraptor. Adults were around 20 feet (6.1 meters) long and around 5 feet tall at the hip. Velociraptor was the size of a turkey.

80. Nose Horn: The Ceratosaurus had a horn on its nose. This was a medium-sized theropod. It was 18 to 23 feet long and had deep jaws that supported very long blade-like teeth.

81. Dinosaur Eyes: Dinosaurs' eye sockets were in different shapes to support better bite strength. Some dinosaurs had forward-facing eyes, which helped them judge distances, and some had eyes on the side of the head for a broader range of vision.

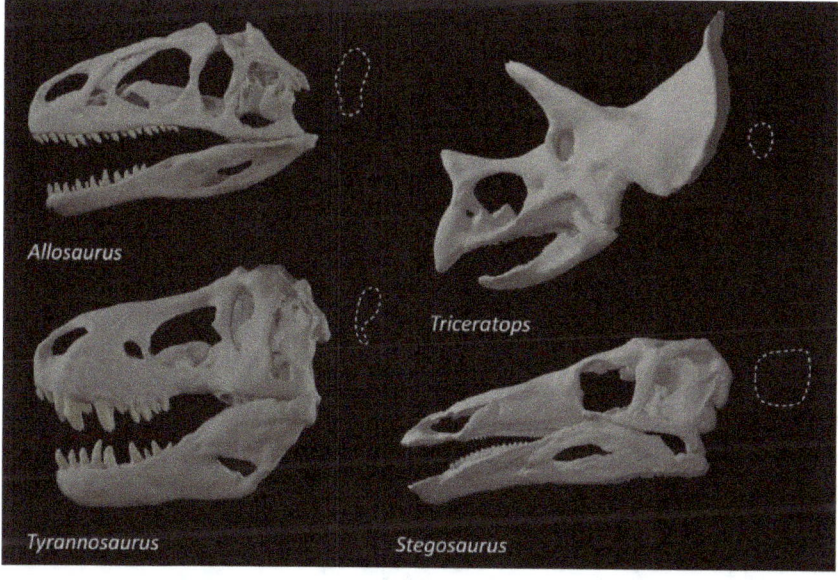

82. Dueling Dinosaurs: Fossils can show evidence of diseases and injuries from battles. Some even show signs of cancer on the bone.

'Fighting dinosaurs' in Tugrugeen Shireh (Gobi Desert, 1971)

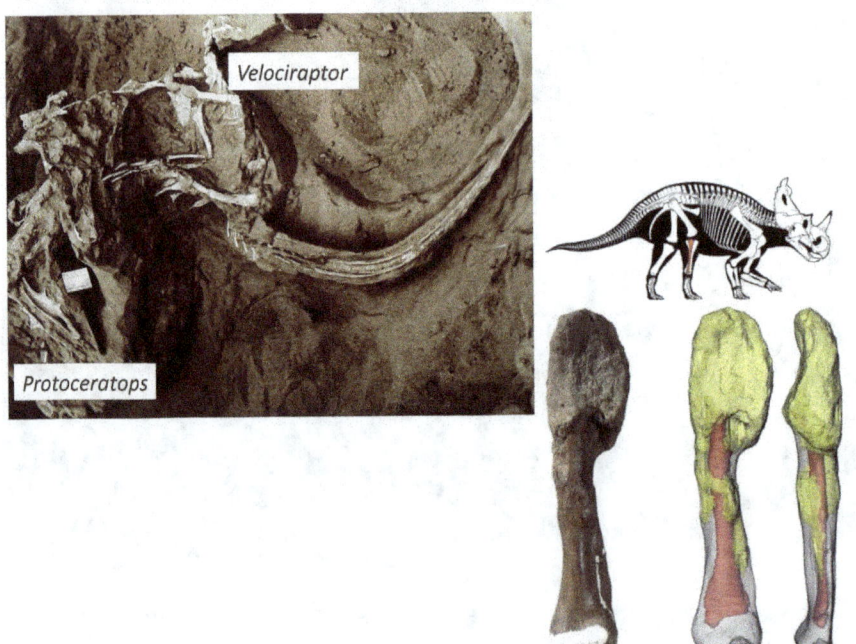

This fossil shows cacer on the legbone of this tricerotops.

83. Thunder Lizard: The name Brontosaurus means "thunder lizard." They earned the name because their steps sounded like thunder when they walked, and they shook the ground. I bet you make noise when you walk through your house, but not that much noise.

Brontosaurus
66 feet (20.1 metres) from head to tail
estimated weight: 28.1–34.5 metric-ton range

3 metres
9 feet

© Encyclopædia Britannica, Inc.

84. Plateosaurus: This dinosaur walked on two legs but could also walk on all fours. It fed from trees high above the ground. You walk on two legs and can walk on all fours, too, if you want to.

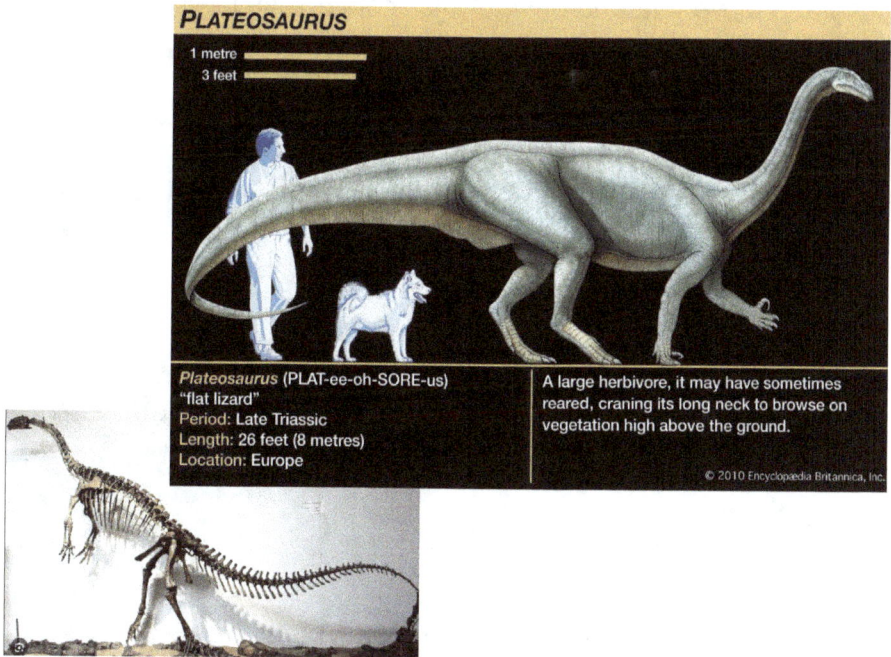

PLATEOSAURUS

1 metre
3 feet

Plateosaurus (PLAT-ee-oh-SORE-us) "flat lizard"
Period: Late Triassic
Length: 26 feet (8 metres)
Location: Europe

A large herbivore, it may have sometimes reared, craning its long neck to browse on vegetation high above the ground.

© 2010 Encyclopædia Britannica, Inc.

85. Water dweller: The Plesiosaurus was a marine reptile, not a dinosaur, with a long neck and flippers. It did live in the time of dinosaurs. This animal even reached 11 feet and more, they were a relatively gentle fish-feeder.

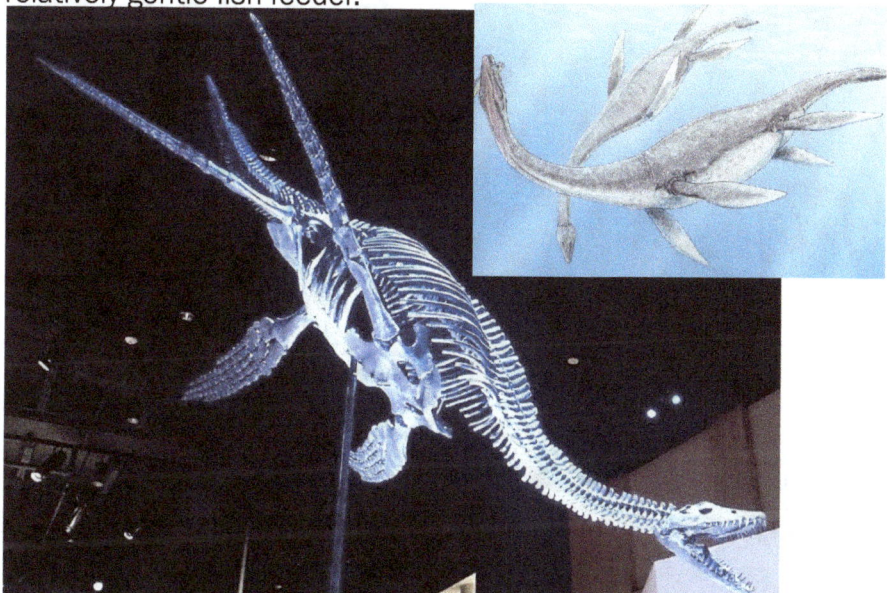

86. Impressive spines: The Amargasaurus had long spines on its neck. The beast grew to a length of 30 to 43 feet long. It weighed about 2.6 to 4 metric tons. It had a long neck and tail with a small head and moved around on four thick column-like legs.

87. Amazing Wingspan: The Quetzalcoatlus was one of the largest flying creatures ever, with a wingspan as large as a small plane. Imagine climbing on his back and flying all around. That would have been amazing!

Art by Mark Witton, from Witton and Naish, 2008.
A GROUP OF QUETZALCOATLUS FORAGES FOR A DINOSAUR DINNER.

88. Fossils: Dinosaur bones can turn into fossils through a process called mineralization. The fossil forms when covered quickly with sediment or mud. After years and years, the seediment erodes away revealing the fossils.

HOW IS A FOSSIL FORMED?

1. Sediment
An animal is buried by sediment, such as volcanic ash or silt, shortly after it dies. Its bones are protected from rotting by the layer of sediment.

2. Layers
More sediment layers accumulate above the animal's remains, and minerals, such as silica (a compound of silicon and oxygen), slowly replace the calcium phosphate in the bones.

3. Movement
Movement of tectonic plates, or giant rock slabs that make up Earth's surface, lifts up the sediments and pushes the fossil closer to the surface.

4. Erosion
Erosion from rain, rivers, and wind wears away the remaining rock layers. Eventually, erosion or people digging for fossils will expose the preserved remains.

89. The Biggest: The Utahraptor is the largest raptor found to date. It had two-inch-long serrated (steak knife) teeth in a foot-and-a-half-long skull. Its claws were up to 10 inches long, with 15-inch sickle-claws on its feet. It was fast and highly agile. It also may have hunted in packs, making it a formidable predator.

Named left to right

1 Microraptor gui
2 Velociraptor mongoliensis
3 Austroraptor cabazai
4 Dromaeosaurus albertensis
5 Utahraptor ostrommaysorum
6 Deinonychus antirrhopus

90. Footprints: A dinosaur's footprints can tell its speed and walking pattern. Did they walk on two legs or four? Did they travel in herds or alone? How fast did they move, and where did they live? Scientists can learn all this and more by studying dinosaurs' tracks.

91. Thorny devil: The Stygimoloch had a dome-shaped head with horns.Some scientists think this dinosaur was just a juvenile Pachycephalosaurus and not a different species. If so, as it grew, it lost its horns and formed domes as they aged. Look up the images. What do you think?

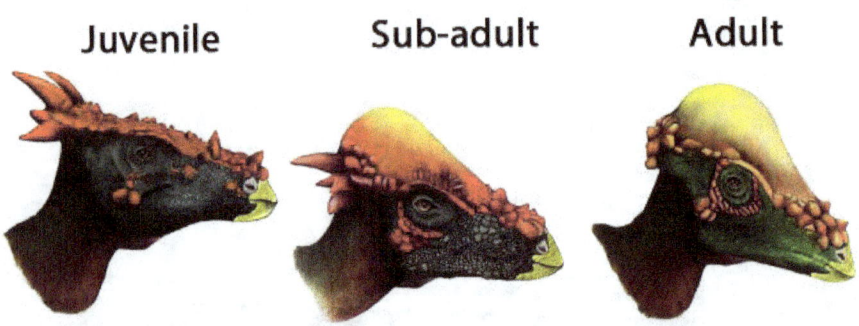

It has been proposed that the cranial ornamentation and skull shape of pachycephalosaurs changes as these animals grow and mature. This can cause confusion when trying to identify species.

Picture Credit: Kari Scannella with additional annotation by Everything Dinosaur

92. Eggs: Dinosaur eggs came in different shapes: round, elongated, or even spiral, in addition to a variety of colors. Each species had eggs with its own characteristics.

Watercolor illustration of different dinosaur eggs. Tyrannosaurus and Diplodocus shown above.

93. Arm and Claws: The T. rex's two-fingered hand was still powerful and could grasp prey. Having two fingers instead of three gave each claw twice as much slashing power, and they were more stable at the wrist. It is believed they could pick up about 400 pounds even if they were short. They could definitely pick up you or me! Wow.

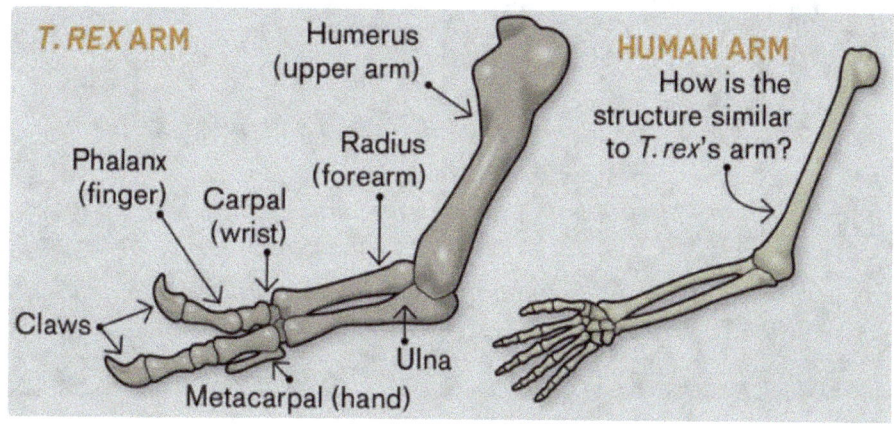

94. New Discoveries: The discovery of fossils is a fascinating window into the past, particularly those that have been unearthed with incredibly well-preserved features. Among these remarkable finds, some fossils stand out for having retained not just their skeletal structure but also more delicate aspects such as feathers, skin, and, in a few extraordinary cases, even the colors of these ancient creatures.

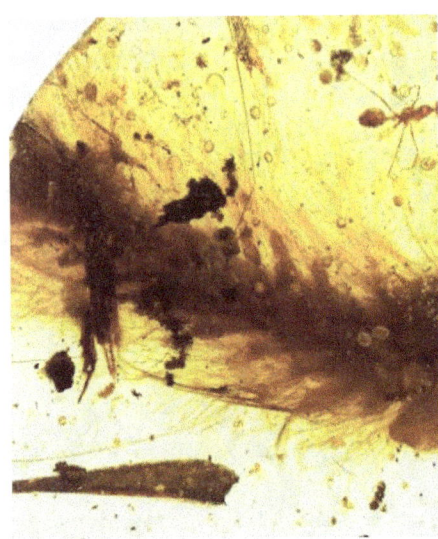

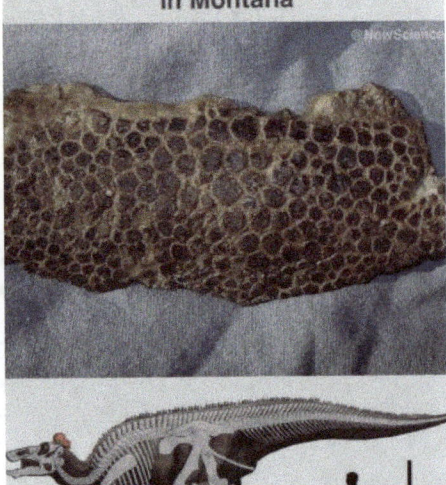

This is a piece of fossilised dinosaur skin of an Edmontosaurus, discovered in Montana

95. Camoflage: Scientists believe that some dinosaurs could change color, similar to chameleons. An example is the find of a fossil in China of the Psittacosaurus. Psittacosaurus means parrot lizard because it had a parrot like beak. Melanosomes, cells that produce and store melanin, found in the fossils would allow for changing colors.

96. Disco Dinosaurs: Scientists think that some dinosaurs may have danced to attract a mate. They would have turned, twisted, and kicked to impress females. Evidence comes from large scraped marks left by dinosaurs in what we know now as Colorado. This behavior is known as "scraping".

97. Jurassic Park or Real: Dilophosaurus means "double-crested lizard. In the real world, this fascinating dinosaur from the early Jurassic period was quite different from its depiction in the popular movie "Jurassic Park." In reality, there's no scientific evidence to suggest that this creature had a frill around its neck or the ability to spit venom, two features that made it quite memorable in the film.

98. Not a dinosaur: Ichthyosaurs were marine reptiles that looked like dolphins. they are not dinosaurs but lived at the same period of time.

99. Babies: Most dinosaurs laid eggs, if not all of them. Some built large nests to keep their babies together. Baby dinosaurs grew very fast; many reached full size when they were seven or eight. You will not be fully grown for 18 to 20 years old. The largest dinosaurs may have lived to be almost 100 years. That is really old, isn't it?

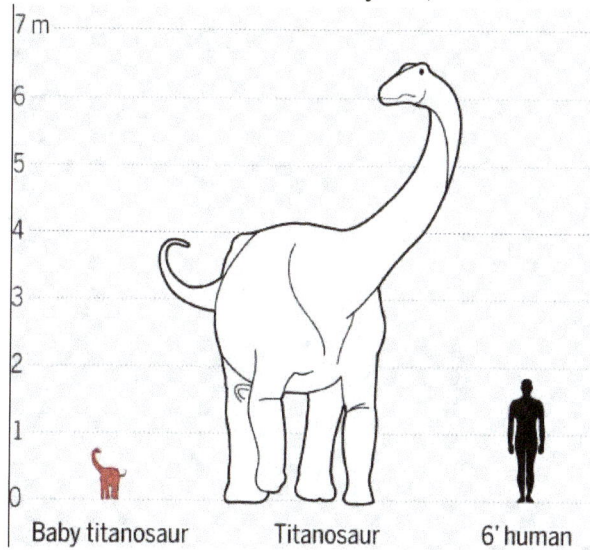

Baby titanosaur   Titanosaur   6' human

100. One of the First: The Coelophysis was one of the earliest known dinosaurs. This dinosaur was small and slender and grew up to almost ten feet long. It walked on two legs and has been found mostly in the southwestern United States. The name Coelophysis comes from a Greek word meaning "hollow form".

Coelophysis compared to a 1.8 meter tall person.

DIMENSIONS IN METERS
©www.prehistoric-wildlife.com 29-02-2016
ILLUSTRATION BY DARREN PEPPER

101. The Archaeopteryx is an ancient bird-like dinosaur that had feathers. They were more like dinosaurs than birds, having sharp teeth, three fingers with claws and a long bony tail. They had toes with a killing claw and were thought to be warm blooded.

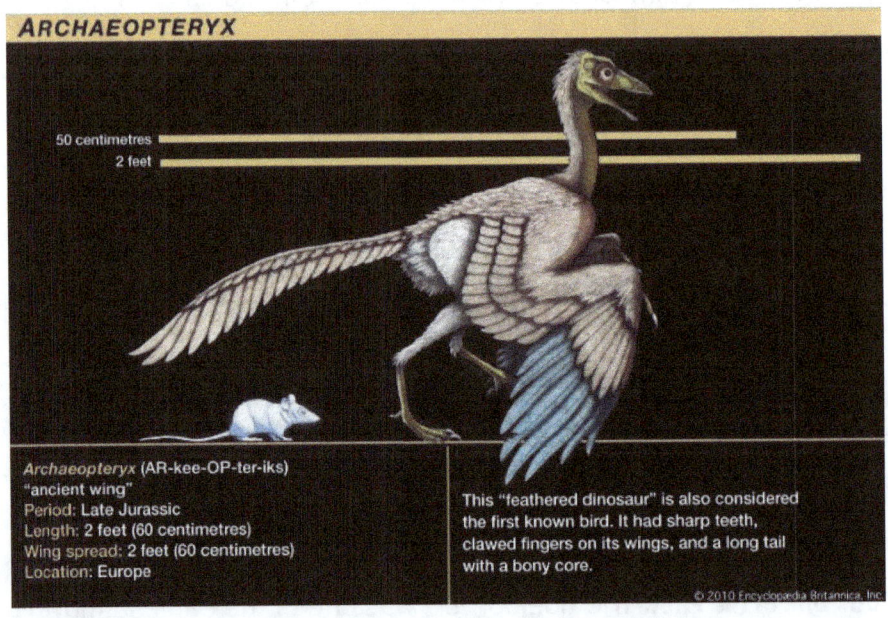

ARCHAEOPTERYX

50 centimetres
2 feet

Archaeopteryx (AR-kee-OP-ter-iks)
"ancient wing"
Period: Late Jurassic
Length: 2 feet (60 centimetres)
Wing spread: 2 feet (60 centimetres)
Location: Europe

This "feathered dinosaur" is also considered the first known bird. It had sharp teeth, clawed fingers on its wings, and a long tail with a bony core.

© 2010 Encyclopædia Britannica, Inc.

ACTIVITY PAGES

See the link at the end of the book to print out full size copies of these coloring and activity pagesand more. Print them in sizes you like and use them as many times as you like!

What color do you want this Allosaurus to be?

Write your dinosaur story here or make a list of your favorite dinosaurs.

Dinosaur

Tyrannosaurus

Match the Picture

Help Mama Dinosaur to find her baby

Dinosaur

Brachiosaurus

CROSSWORD

Name the Dinosaurs

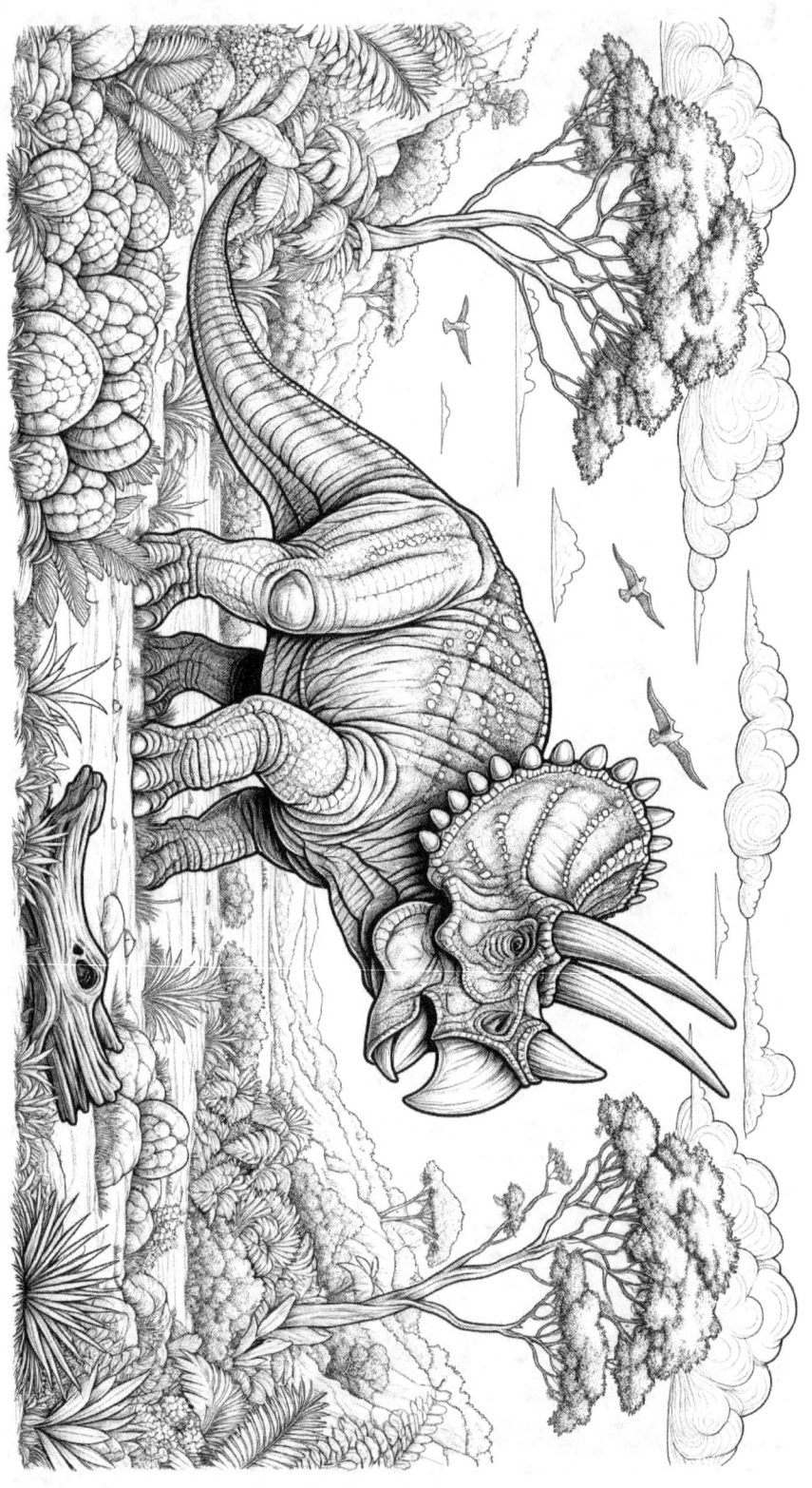

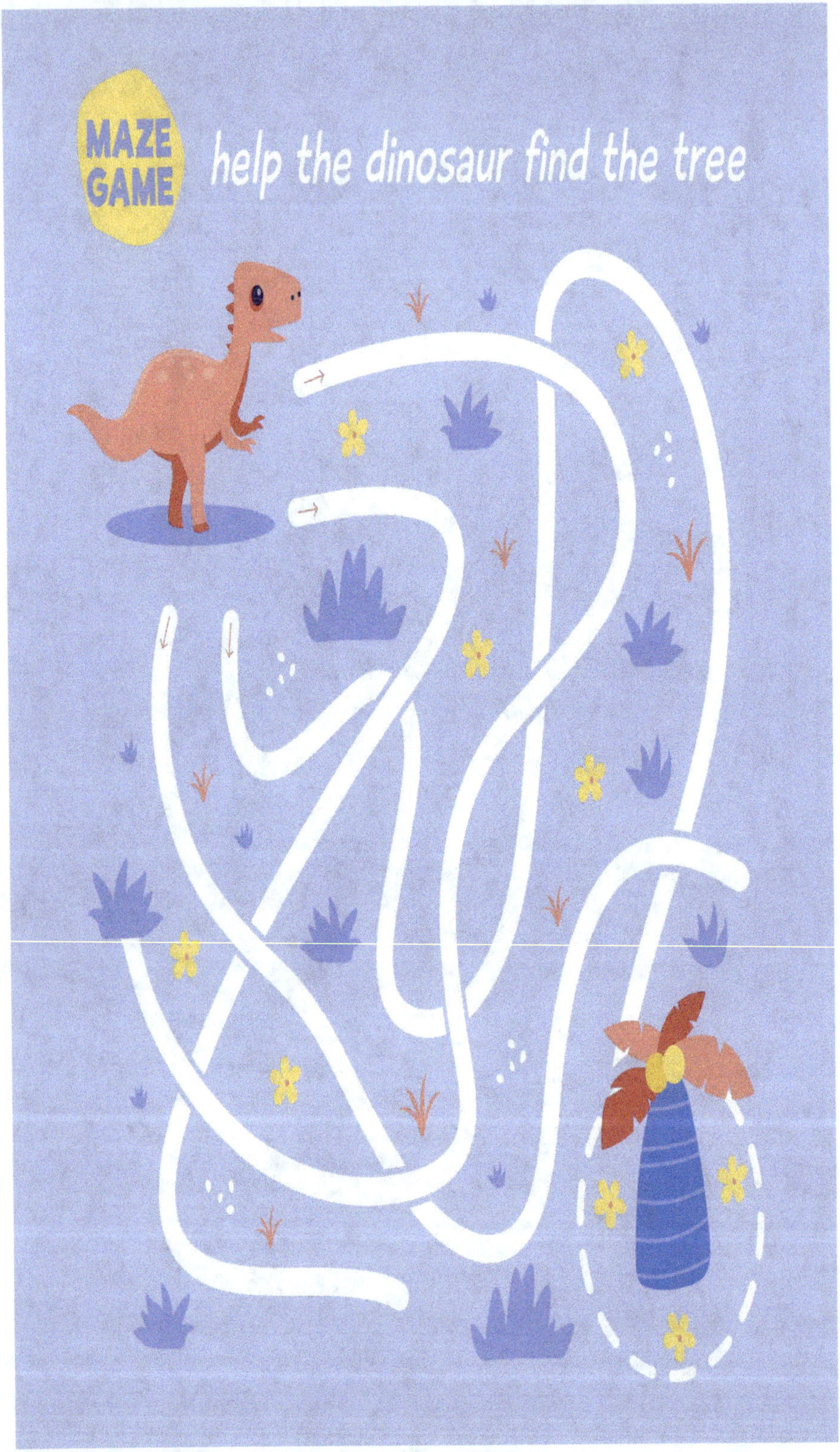
MAZE GAME
help the dinosaur find the tree

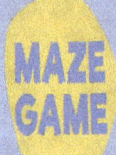

CONCLUSION

Wow, what an adventure we've been on, fellow dino-enthusiasts!

I hope you enjoyed this thrilling journey through the ancient world of dinosaurs! It's been an extraordinary adventure, filled with awe-inspiring moments and fascinating discoveries.

Imagine the vast landscapes of prehistoric Earth, where these incredible creatures roamed freely. From the majestic Apatosaurus, stretching its long neck towards the heavens, to the swift Velociraptor, dashing through the underbrush. We've encountered dinosaurs of all shapes and sizes, each with their own unique characteristics and habits. These magnificent beasts not only dominated the land but also the air and seas, showcasing the incredible diversity of life that once existed on our planet.

It's amazing to think that what we know about dinosaurs today comes from the remnants they left behind millions of years ago. Fossilized bones, footprints, and even traces of their feathers paint a picture of a world very different from ours. But this picture is incomplete. There are still countless secrets buried deep within the Earth, waiting for us to find them. New species of dinosaurs are discovered regularly, adding more pieces to the puzzle of our planet's history.

And let's not forget the unsung heroes of this story - the paleontologists. These dedicated scientists work tirelessly, excavating and studying fossils to unlock the mysteries of the past. Their discoveries not only help us understand the dinosaurs but also how life on Earth has evolved over time.

You're stepping into a world brimming with possibilities. Maybe one day, you'll be the one uncovering a new dinosaur species or solving an ancient mystery. Keep asking questions, exploring, and dreaming. After all, today's young dreamers are tomorrow's adventurers and scientists. Who knows what discoveries await you in the future? So, keep that spark of curiosity alive. Let it guide you on your journey through life, and always remember the incredible world of dinosaurs - a testament to the wonders of our natural world and the endless pursuit of knowledge. Keep exploring, keep learning, and who knows, maybe you'll write the next chapter in this never-ending story of discovery!

This adventure might seem like it's coming to an end as you close the book, but in reality, it's just the beginning. Every page you turned opened a new doorway to knowledge and imagination. Remember, each dinosaur started as a small egg and grew into something extraordinary. Just like them, your curiosity and passion for learning can grow into something amazing. Who knows what you might discover!

REFERENCES

Natural History Museum Website: https://www.nhm.ac.uk/ Publisher: The Trustees of the Natural History Museum, London Date Accessed: [October 2023]

American Museum of Natural History. (October, 2023). (Paleontology Section). https://www.amnh.org/

Title: Britannica Website: https://www.britannica.com/ Publisher: Encyclopædia Britannica, Inc. Date Accessed: [October 2023]

Prehistoric Wildlife. (n.d.). Micropachycephalosaurus. Retrieved November 9, 2023, from https://www.prehistoric-wildlife.com/species/m/micropachycephalosaurus.html

Smithsonian National Museum of Natural History. (2023). [3D Model of Tyrannosaurus Rex]. Retrieved from https://collections.nmnh.si.edu/search/prehistoric-life/Tyrannosaurus-rex

KRISTIN ROMEY, November 2023. Feathered dinosaur tail discovered in amber: New find from Myanmar includes insects, leaves, and lizard-like creatures preserved for 99 million years. National Geographic. Retrieved from https://www.nationalgeographic.com/animals/article/feathered-dinosaur-tail-amber-theropod-myanmar-burma-cretaceous. Fact #94.

Dear Reader,

First and foremost, thank you for taking the time to delve into the pages of this book. Your engagement and enthusiasm are invaluable to me.

As an independent author, every aspect of bringing this story to life, from the first word on the page to the moment it reaches readers like you, is a journey filled with passion and dedication. Your insights and reactions play a crucial role in this journey.

Now that you've experienced the world and the creatures you and I have learned about, I have a small but significant request. I would be deeply grateful if you could take a moment to leave a review.

https://www.amazon.com/Fun-Facts-Kids-Patricia-Williamson-ebook/ dp/B0CNSMCM86/ref=sr_1_5?crid=1GDTCTLKVAQRR&keywords=Fun+- Facts+for+Kids+Dinosaurs&qid=1701102214&s=books&sprefix=fun+- facts+for+kids+dinosaurs%2Cstripbooks%2C79&sr=1-5

Your honest thoughts and reflections not only help other readers discover this story but also provide me with essential feedback that shapes my future work.

Your review can be as brief or as detailed as you like. What mattered most to you? Which characters resonated with you? Every word of your review is a stepping stone in my writing journey and contributes to the larger community of book lovers and independent authors.

Thank you once again for your time and for being part of this book's journey. Your support is a beacon that guides and inspires.

With heartfelt appreciation,

Patricia Williamson

P.S. Remember your words and opinions continue to make a difference. Happy reading and reviewing!

This is the URL and QR code to the printable activity pages.

Have Fun!

https://bit.ly/DinosaurActivities https://bit.ly/DinosaurActivities